Human Anatomy Coloring Book

50 Illustrations to Learn Anatomy & Physiology with Easy Coloring and Interactive Study: Made for Adults, Medical Students, and Healthcare Professionals!

Medical Pass

Table of Contents

$100+ FREE BONUSES

**100 Human Anatomy
Flash Cards**

**Color-Coding Anatomy
Cheatsheet**

**Human Anatomy
Quizzes**

Scan QR code to claim
your bonuses

— OR —

visit bit.ly/4exyDHg

Introduction

Welcome to the Human Anatomy Coloring Book! This comprehensive guide offers 50 detailed illustrations designed to help you learn anatomy and physiology through the engaging process of coloring.

Whether you're an adult looking to deepen your understanding of the human body, a medical student preparing for exams, or a healthcare professional seeking a fun and interactive study tool, this book is tailor-made for you.

How to Use:

1. **Choose your topic:** Begin by starting with the first illustration or selecting a specific area of anatomy that you'd like to explore. With 50 illustrations covering various organ systems and anatomical regions, you have a wide range of options.

2. **Color and Learn:** Once you've chosen a section to study, grab your coloring pencils and dive in! Each illustration is carefully crafted to highlight key anatomical structures, with numbered labels corresponding to a detailed list of structures provided alongside. Color each structure according to its corresponding number, allowing you to reinforce your knowledge as you go.

3. **Engage with Interesting Facts:** As you color, take a moment to read the "Interesting Facts" section provided for each illustration. These tidbits of information offer valuable insights and trivia about the structures you're coloring, enriching your learning experience.

4. **Review:** After completing an illustration, take a moment to review the labeled structures and reflect on what you've learned.

So grab your coloring pencils and embark on a journey of discovery through the fascinating world of anatomy and physiology!

The Human Cell

The human cell is the basic structural and functional unit of life. From simple bacteria to complex multicellular organisms like humans, all living organisms are composed of cells. This section provides a detailed exploration of the cell's fundamental structures. Let's start exploring them now.

1. Cell Membrane

2. Cytoplasm

3. Nucleus

4. Endoplasmic Reticulum

5. Ribosomes

6. Golgi Apparatus

7. Mitochondria

8. Lysosomes

9. Peroxisomes

10. Centrosome

11. Microtubules

12. Microfilaments

13. Intermediate Filaments

14. Flagella

15. Cilia

16. Nuclear Envelope

17. Nucleolus

18. Chromatin

Interesting Fact

Did you know that the average human body contains trillions of cells, each performing specialized functions essential for survival? Despite their microscopic size, cells are remarkably complex and diverse, forming the building blocks of tissues, organs, and organ systems throughout the body.

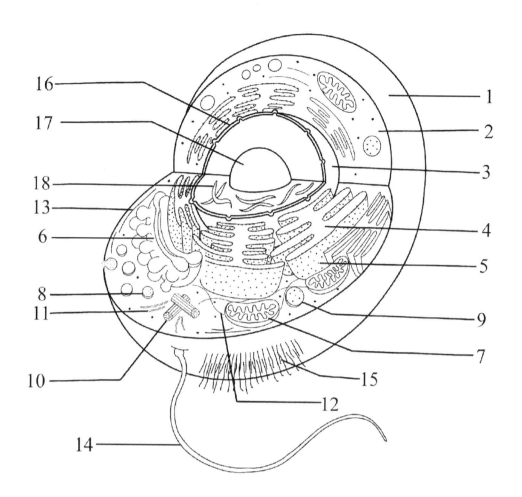

16 ———————————————— 1

17 ———————————————— 2

18 ———————————————— 3

13 ————————————————
6 ———————————————— 4

8 ———————————————— 5

11 ————————————————

10 ———————————————— 9

14 ———————————————— 7

15

12

Types of cells

The human body comprises various types of cells, each with its unique structure and function. Understanding the diversity of human cells is essential for grasping the complexity of human anatomy and physiology. Below is a list of different types of human cells found throughout the body.

1. Stem Cell

2. Epithelial Cell

3. Red blood cells (Erythrocyte)

4. White blood cells (Leukocyte)

5. Platelet (Thrombocyte)

6. Nerve cell

7. Adipocytes (Adipose or fat cells)

8. Sperm

9. Ovum (egg)

10. Osteocyte (Bone Cell)

11. Chondrocyte (Cartilage Cell)

12. Fibroblast (Connective tissue cells)

Interesting Fact

The human body is a marvel of cellular diversity, with over 200 different types of cells performing specialized functions. Each cell type contributes uniquely to the body's complex ecosystem, from neurons transmitting electrical signals to muscle cells, which contract for movement.

Types of cell

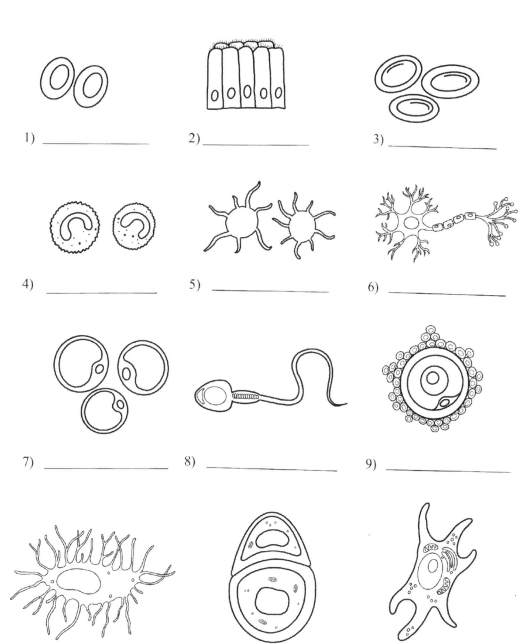

1) _____

2) _____

3) _____

4) _____

5) _____

6) _____

7) _____

8) _____

9) _____

10) _____

11) _____

12) _____

Blood Cells and Immune Cells

Dive into the vibrant world of blood and immune cells, essential components of the body's defense and transport systems. Explore the intricate workings of these vital cell types, from oxygen-carrying red blood cells to infection-fighting white blood cells.

1. Red Blood Cell (Erythrocyte)

2. Neutrophils

3. Eosinophil

4. Basophils

5. Monocyte

6. Platelet (Thrombocyte)

7. Macrophage

8. B-lymphocyte

9. T-lymphocyte

10. Natural Killer (NK) Cells

11. Dendritic Cells

12. Mast Cells

Interesting Fact

Blood and immune cells work together seamlessly to protect the body from infections and maintain homeostasis. Despite their distinct roles, they collaborate harmoniously to ensure the body's survival and resilience.

Blood cells and Immune cells

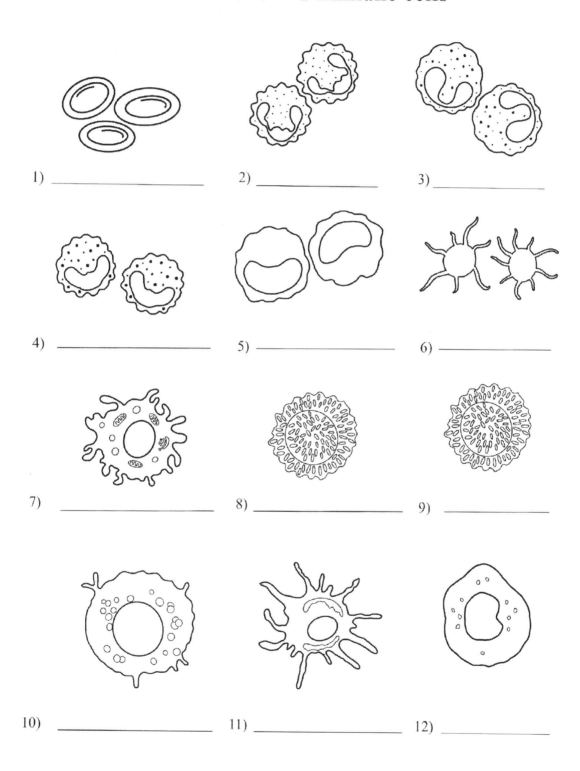

1) _____

2) _____

3) _____

4) _____

5) _____

6) _____

7) _____

8) _____

9) _____

10) _____

11) _____

12) _____

Types of Muscles

Welcome to an insightful exploration of the diverse muscle types that underpin human anatomy. From the precision-driven skeletal muscles facilitating our every movement to the tireless pulsations of cardiac muscle ensuring our heart's rhythmic beat, the muscles can be divided into three types. Let's explore them now.

1. Skeletal Muscle

2. Cardiac Muscle

3. Smooth Muscle

Now, let's explore the structure of the muscle fiber of a skeletal muscle.

1. Bone

2. Tendon

3. Epimysium

4. Perimysium

5. Endomysium

6. Myofibril

7. Sarcolemma

8. Fascicle

Interesting Fact

Muscle tissue accounts for approximately 40% of total body weight in humans. While skeletal muscles facilitate voluntary movements like walking and lifting, smooth muscles control involuntary actions such as digestion and breathing, showcasing the versatility of muscle types in maintaining bodily functions.

Types of Muscles

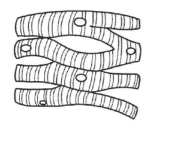 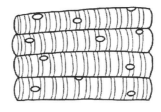 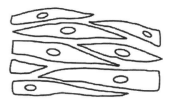

1) _____ 2) _____ 3) _____

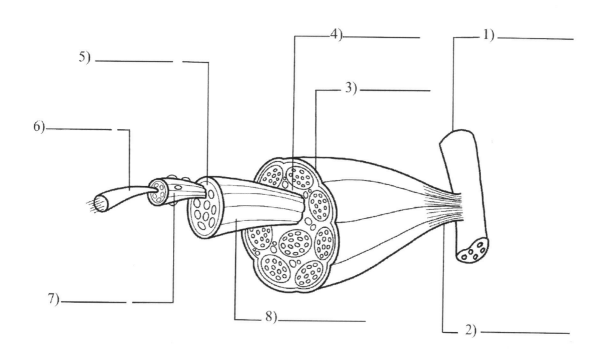

5) _____

6) _____

7) _____

4) _____

3) _____

1) _____

8) _____

2) _____

Skeletal System

Let's step into the human body's structural framework as we explore the parts of the Skeletal System. This system provides support and protection and serves as the anchor for muscles, facilitates movement, and houses the body's vital organs. The system is divided into two main parts: the axial and appendicular skeletons. Let's explore the main parts of the two systems.

1. Skull

2. Ossicles

3. Hyoid Bone

4. Vertebral column

5. Rib Cage

6. Shoulder Girdle

7. Arm

8. Hand

9. Pelvic Girdle

10. Leg

11. Patella

12. Foot

Interesting Fact

Did you know the human skeleton is a dynamic and adaptable structure, constantly remodeling and reshaping itself throughout life? From infancy to adulthood, bone tissue undergoes a continuous renewal process, allowing our skeletons to adapt to changes in activity levels, environment, and physiological demands. This remarkable capacity for regeneration underscores the skeletal system's essential role in our ongoing growth and development.

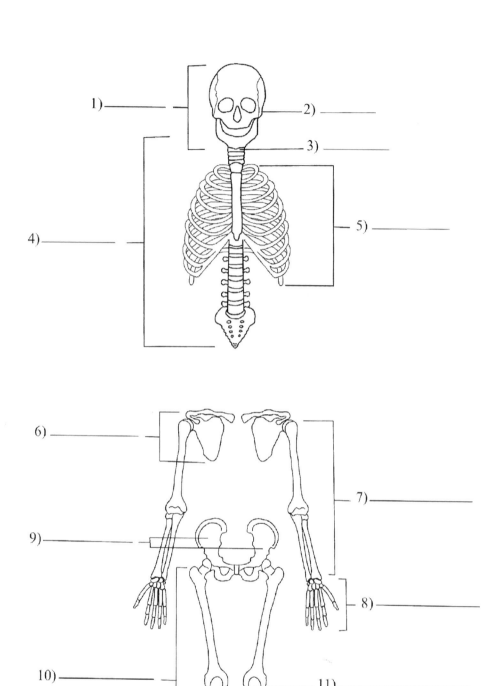

1) _____

2) _____

3) _____

4) _____

5) _____

6) _____

7) _____

8) _____

9) _____

10) _____

11) _____

12) _____

Skull and Facial Bones

The skull comprises two types of bones: facial and cranial bones. These form the intricate framework that encases and protects the brain while defining the human face's unique features. Let's delve into the fascinating anatomy of these structures.

1. Frontal Bone

2. Parietal Bone

3. Temporal Bone

4. Occipital Bone

5. Sphenoid Bone

6. Ethmoid Bone

7. Nasal Bone

8. Maxilla

9. Zygomatic Bone

10. Lacrimal Bone

11. Palatine Bone

12. Inferior Nasal Conchae

13. Vomer

14. Mandible

15. Glabella

16. Orbit

17. Infraorbital foramen

18. Mental foramen

Sutures in the skull

1. Sagittal suture

2. Coronal suture

3. Rhomboid suture

4. Squamous suture

Interesting Fact

The human skull consists of 22 bones, including 14 facial and 8 cranial bones. Remarkably, these bones are not fully fused at birth, allowing for the flexibility to accommodate rapid brain growth during infancy.

Sutures of the Skull

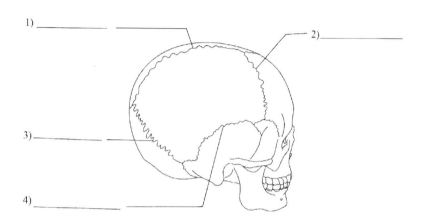

1) _____

2) _____

3) _____

4) _____

Skull and Facial bones

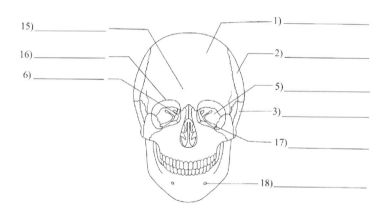

15) _____

16) _____

6) _____

1) _____

2) _____

5) _____

3) _____

17) _____

18) _____

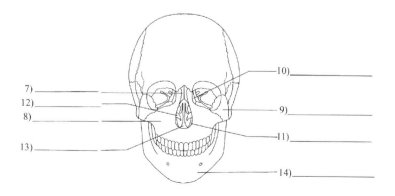

7) _____

12) _____

8) _____

13) _____

10) _____

9) _____

11) _____

14) _____

Vertebral Column and Rib Cage

The vertebral column and rib cage form the core of the human skeletal system, providing essential support and protection to vital organs such as the spinal cord and thoracic cavity. Together, these interconnected structures enable flexibility, stability, and respiration.

1. Cervical Vertebrae (C1-C7)

2. Thoracic Vertebrae (T1-T12)

3. Lumbar Vertebrae (L1-L5)

4. Sacral Vertebrae

5. Coccygeal Vertebrae

6. Manubrium of Sternum

7. Body of Sternum

8. Xiphoid process

9. True Ribs (Ribs 1-7)

10. False Ribs (Ribs 8-12)

11. Floating Ribs (Ribs 11-12)

12. Costal Cartilage

Vertebral Column and Rib Cage

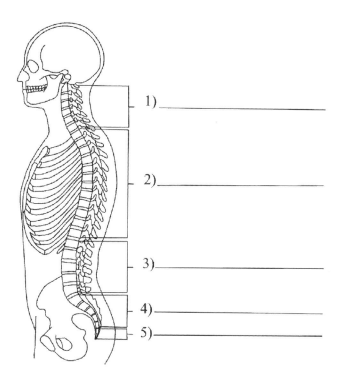

1) _____

2) _____

3) _____

4) _____

5) _____

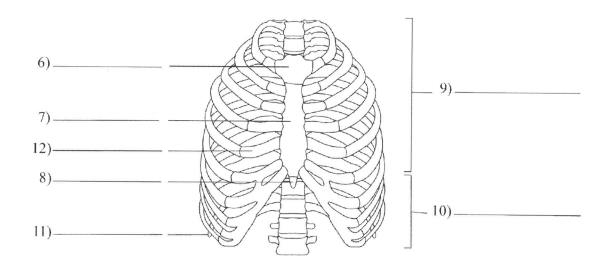

6) _____

7) _____

12) _____

8) _____

11) _____

9) _____

10) _____

Upper Limb and Shoulder Girdle

The upper limb and shoulder girdle constitute the intricate system responsible for the versatile movements of the arms and hands. These provide mobility and stability, facilitating everyday activities and dynamic athletic performances.

1. Clavicle (Collarbone)

2. Scapula (Shoulder Blade)

3. Humerus

4. Radius

5. Ulna

6. Carpals (Wrist Bones)

7. Metacarpals (Palm Bones)

8. Phalanges (Finger Bones)

9. Elbow joint

10. Acromion

The Hand

1. Scaphoid

2. Lunate

3. Triquetrum (Triquetral)

4. Pisiform

5. Hamate

6. Capitate

7. Trapezoid

8. Trapezium

9. Metacarpal bone

10. Proximal phalange

11. Distal phalange

12. Middle phalange

13. Sesamoid bone

Interesting Fact

The clavicle, often called the "collarbone," is the only long bone in the human body that lies horizontally, connecting the sternum to the scapula. This distinctive feature provides stability to the shoulder joint and allows for greater flexibility and range of motion in the upper limb.

Upper Limb and Shoulder Girdle

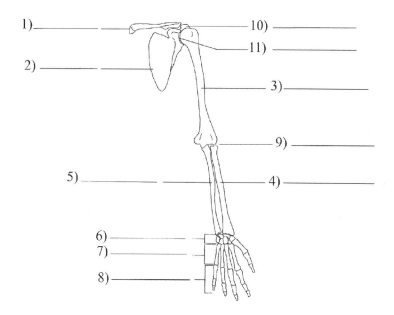

1) _____

2) _____

5) _____

6) _____

7) _____

8) _____

10) _____

11) _____

3) _____

9) _____

4) _____

The Hand

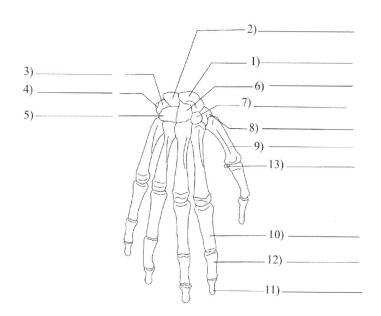

2) _____

3) _____

4) _____

5) _____

1) _____

6) _____

7) _____

8) _____

9) _____

13) _____

10) _____

12) _____

11) _____

Lower Limb and Pelvic Girdle

The lower limb is the foundation for human movement, supporting our body weight and facilitating activities such as walking, running, and jumping. Comprised of a complex network of bones, muscles, and joints, this vital anatomical region plays a crucial role in our mobility and daily functioning. Let's learn the anatomical structures that form the lower limb and pelvic girdle.

1. Pelvic Girdle

2. Femur

3. Patella (Kneecap)

4. Tibia

5. Fibula

6. Tarsal Bones

7. Metatarsal Bones

8. Phalanges

Pelvic Girdle

1. Ilium

2. Ischium

3. Pubis

4. Pubic symphysis

Interesting Fact

Among the bones of the lower limb, the femur, or thigh bone, stands out as the longest and strongest bone in the human body. Its robust structure not only provides essential support but also withstands tremendous forces during weight-bearing activities, showcasing the remarkable resilience of our skeletal system.

Lower limb and pelvic girdle

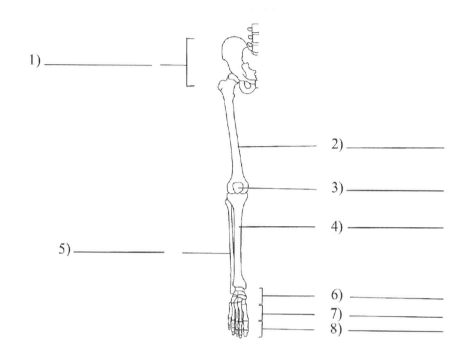

1) _____

2) _____

3) _____

4) _____

5) _____

6) _____

7) _____

8) _____

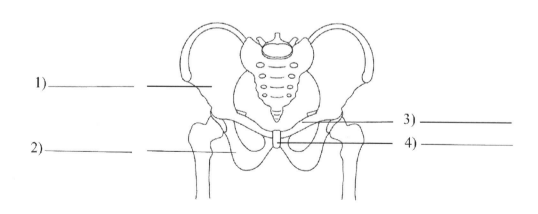

1) _____

2) _____

3) _____

4) _____

Joints

Joints play a pivotal role in human anatomy, serving as the connectors between bones and enabling movement and flexibility in the body. From the robust ball-and-socket joints facilitating wide-ranging movements to the intricate hinge joints facilitating precise actions, let's explore the different types of joints within the human body.

1. Ball-and-Socket Joint

2. Hinge Joint

3. Pivot Joint

4. Planar or Gliding Joint

5. Saddle Joint

6. Condyloid Joint

Interesting Fact

The human body contains over 350 joints, ranging from highly mobile synovial to rigid fibrous joints. Despite their structural differences, all joints share the common goal of facilitating movement and ensuring the integrity of the musculoskeletal system.

Joints

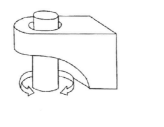

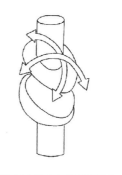

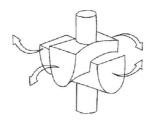

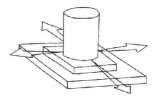

The knee joint

The knee joint is a remarkable and complex structure that plays a crucial role in weight-bearing, locomotion, and stability. Composed of several interconnected bones, ligaments, and cartilage, it undergoes tremendous stress daily, making it susceptible to injury and wear over time. Let's explore the anatomy of this essential joint and unravel its intricate workings.

1. Femur

2. Tibia

3. Fibula

4. Patella

5. Articular Cartilage

6. Meniscus

7. Anterior Cruciate Ligament (ACL)

8. Posterior Cruciate Ligament (PCL)

9. Medial Collateral Ligament (MCL)

10. Lateral Collateral Ligament (LCL)

11. Patellar Ligament

Synovial joint in the knee

1. Bone

2. Articular Cartilage

3. Articular Capsule

4. Synovial Membrane

5. Synovial Fluid

Interesting Fact

The knee joint is the largest and most complex joint in the human body. Despite its size, it relies heavily on surrounding ligaments and muscles for stability and support. Additionally, the knee joint is uniquely designed to withstand significant forces while maintaining a delicate balance between mobility and strength, highlighting its remarkable adaptability and resilience.

The knee joint

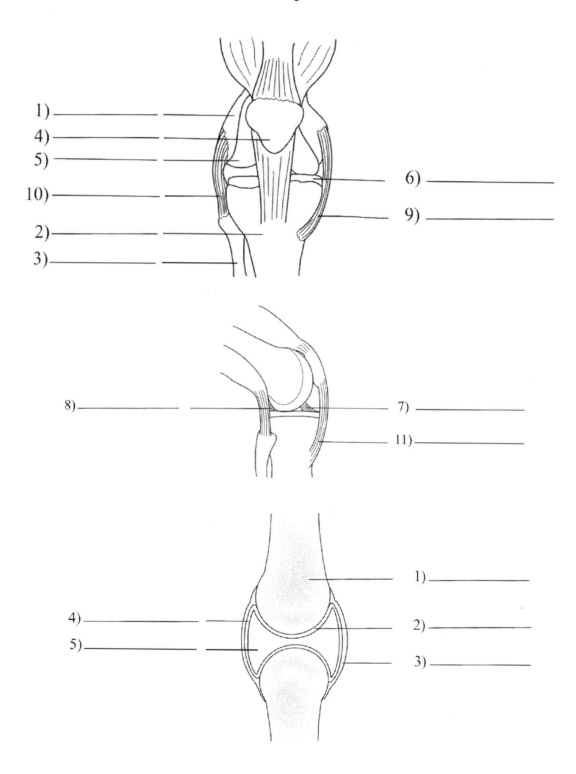

1)

4)

5)

10)

2)

3)

6)

9)

8)

7)

11)

1)

4)

2)

5)

3)

Foot Bones and Ankle joint

The knee joint is a remarkable and complex structure that plays a crucial role in weight-bearing, locomotion, and stability. Composed of several interconnected bones, ligaments, and cartilage, it undergoes tremendous stress daily, making it susceptible to injury and wear over time. Let's explore the anatomy of this essential joint and unravel its intricate workings.

The Foot bones

1. Calcaneus
2. Talus
3. Navicular
4. Cuboid
5. Cuneiform
6. Metatarsal Bones
7. Proximal phalange
8. Middle phalange
9. Distal phalange

The Ankle Joint

1. Tibia
2. Fibula
3. Talus
4. Calcaneus
5. Posterior Talofibular ligament
6. Anterior Talofibular ligament
7. Calcaneofibular ligament

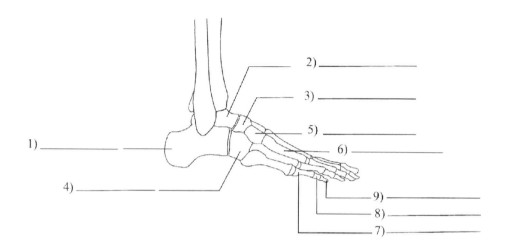

1) _____

2) _____

3) _____

4) _____

5) _____

6) _____

7) _____

8) _____

9) _____

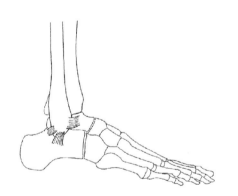

The Temporomandibular joint

The temporomandibular joint (TMJ) is a unique joint that connects the mandible (lower jaw) to the skull's temporal bone. It allows for movements required for chewing, speaking, and facial expressions.

1. Temporal bone

2. Mandible

3. Lateral pterygoid muscle

4. External auditory meatus

5. Tympanic plate

6. Mandibular condyle

7. Temporomandibular joint capsule

8. Articular disc

9. Superior joint cavity

10. Inferior joint cavity

Interesting Fact

The TMJ is one of the most frequently used joints in the human body. Its movements occur thousands of times daily during activities like eating and talking.

Temporomandibular Joint

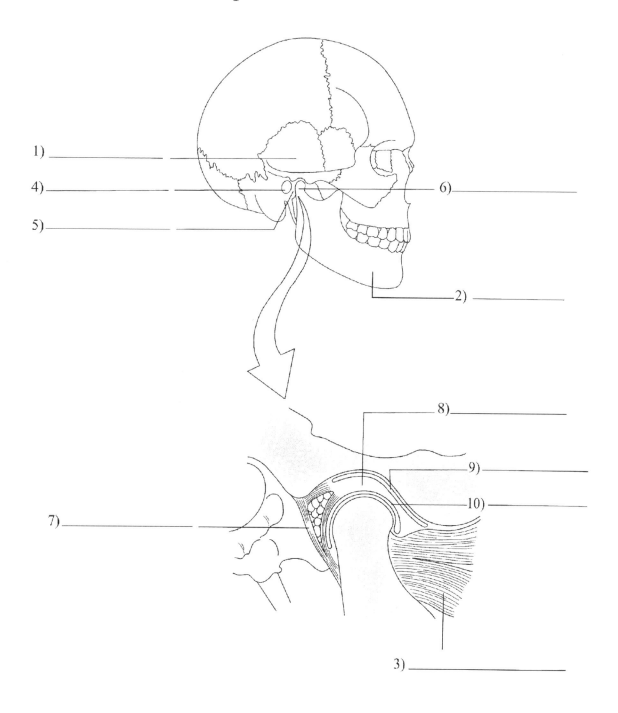

1) _____

4) _____

5) _____

6) _____

2) _____

8) _____

9) _____

10) _____

7) _____

3) _____

The Vertebra

The vertebral column, also known as the spine, is a vital anatomical structure that provides support, protection, and flexibility to the human body. Composed of a series of interconnected vertebrae, this intricate system not only houses and protects the spinal cord but also facilitates various movements and postures. Let's delve into the anatomy of the vertebrae to uncover each type's unique features and functions.

1. Vertebral Foramen
2. Transverse foramen
3. Transverse process
4. Superior articular facet
5. Anterior arch
6. Posterior arch
7. Dens
8. Body
9. Pedicle
10. Lamina
11. Spinous process
12. Anterior tubercle
13. Posterior tubercle
14. Odontoid process

Interesting Fact

The spine is at its maximum length in the morning and shrinks slightly by the end of the day due to compression from daily activities and gravitational forces. This phenomenon can lead to a height difference of up to 1 inch between morning and night, highlighting the dynamic nature of the vertebral column.

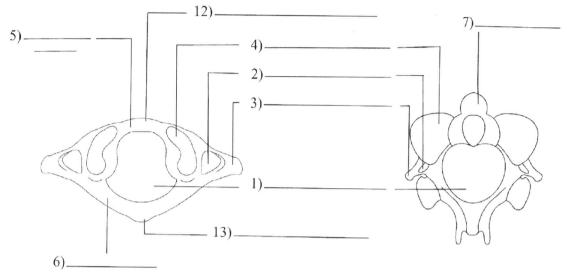

Cervical Vertebrae

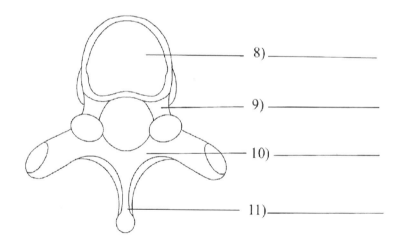

Thoracic Vertebrae

Muscles of Head and Neck

The head and neck muscles play crucial roles in various functions, including facial expression, mastication (chewing), swallowing, and maintaining posture. They are also involved in intricate movements necessary for speech and communication. Let's examine these muscles.

1. Temporalis
2. Occipitalis
3. Orbicularis oculi
4. Frontalis
5. Nasalis
6. Masseter
7. Buccinator
8. Levator labii superioris
9. Zygomaticus major
10. Zygomaticus minor
11. Orbicularis oris
12. Depressor labii inferioris
13. Depressor anguli oris
14. Sternocleidomastoid
15. Trapezius
16. Omohyoid
17. Sternohyoid
18. Levator scapulae

Interesting Fact

The buccinator muscle, located in the cheek, is responsible for compressing the cheek against the teeth, assisting in actions such as blowing and sucking. Interestingly, it's also known as the "trumpeter's muscle" as it's essential for blowing air into wind instruments.

Muscles of Head & Neck

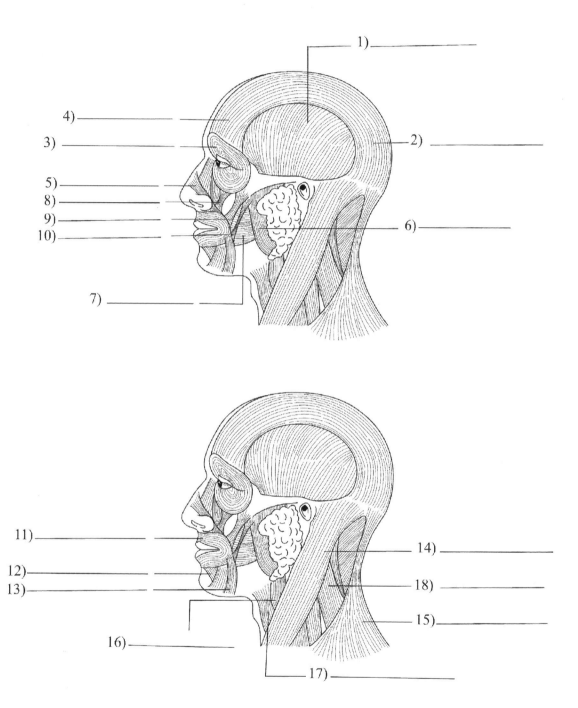

1) _____

4) _____
3) _____
5) _____
8) _____
9) _____
10) _____

2) _____

6) _____

7) _____

11) _____
12) _____
13) _____

14) _____
18) _____
15) _____

16) _____

17) _____

Muscles of the Torso

The torso muscles are essential for respiration, posture, and trunk movement. They also stabilize the spine and play a role in various activities, such as breathing and bending.

1. Pectoralis major

2. Trapezius

3. Latissimus dorsi

4. Serratus anterior

5. Transverse abdominal

6. Internal abdominal oblique

7. External abdominal oblique

8. Rectus abdominis

9. Linea Semilunaris

10. Linea Alba

11. Rhomboid major

12. Rhomboid Minor

13. Spinalis thoracis

14. Longissimus thoracis

15. Iliocostalis

16. Serratus posterior inferior

Interesting Fact

The pectoralis major, commonly known as the pecs, are large muscles in the chest. They play a significant role in movements such as pushing and hugging. Interestingly, the pectoralis major is one of the muscles most commonly developed through weightlifting exercises, contributing to the well-defined chest seen in bodybuilders.

Muscles of the front torso

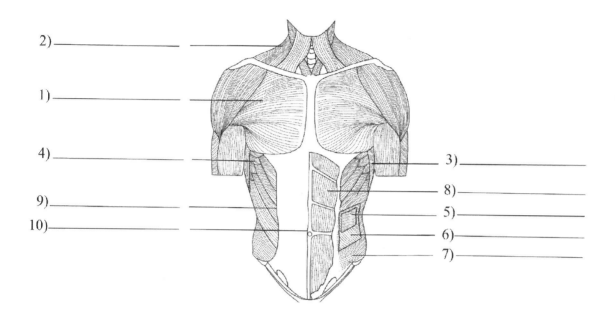

2) —————————————

1) —————————————

4) —————————————

9) —————————————

10) —————————————

3) —————————————

8) —————————————

5) —————————————

6) —————————————

7) —————————————

Muscles of the back torso

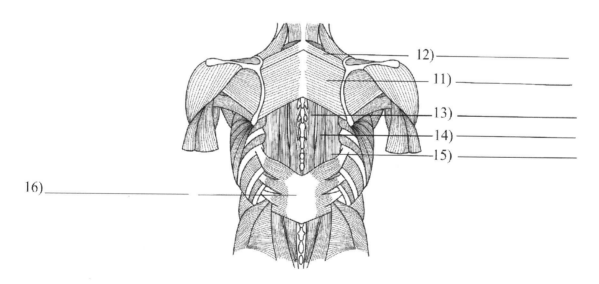

12) —————————————

11) —————————————

13) —————————————

14) —————————————

15) —————————————

16) —————————————

Muscles of the Upper Limb

The upper limb muscles are crucial for performing a wide range of movements, including reaching, grasping, and manipulating objects. These muscles work together with the bones and joints of the arm, forearm, and hand to facilitate various actions, from simple tasks to intricate movements.

1. Deltoid

2. Biceps brachii

3. Brachialis

4. Brachioradialis

5. Triceps brachii

6. Pronator teres

7. Flexor carpi radialis

8. Palmaris longus

9. Flexor carpi ulnaris

10. Flexor digitorum superficialis

11. Flexor digitorum profundus

12. Flexor pollicis longus

13. Extensor carpi radialis longus

14. Extensor carpi radialis brevis

15. Extensor digitorum

16. Extensor digiti minimi

17. Extensor carpi ulnaris

18. Abductor pollicis longus

19. Adductor pollicis longus

20. Extensor pollicis brevis

21. Extensor pollicis longus

22. Extensor indicis

Interesting Fact

The deltoid muscle, named after its triangular shape, is responsible for lifting the arm away from the body, a movement known as abduction. It's one of the primary muscles involved in activities like throwing a ball or reaching for objects overhead.

Muscles of the upper limb

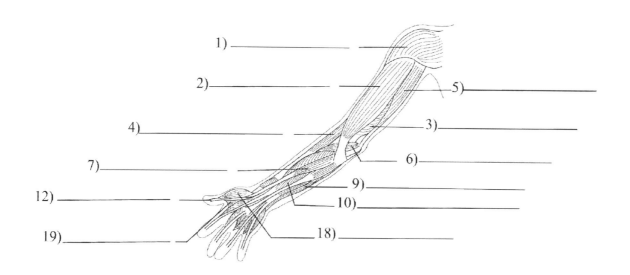

1) _____
2) _____
4) _____
7) _____
12) _____
19) _____
5) _____
3) _____
6) _____
9) _____
10) _____
18) _____

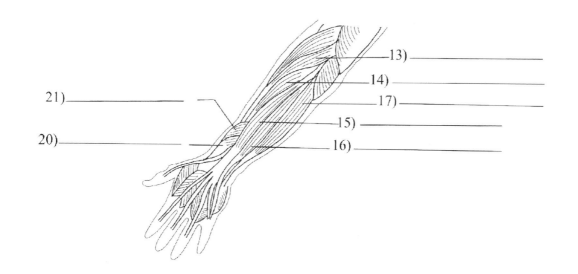

21) _____
20) _____
13) _____
14) _____
17) _____
15) _____
16) _____

Muscles of the Lower Limb

The lower limb muscles are essential for various activities such as walking, running, and maintaining balance. These muscles work together to support the body's weight, stabilize the hip, knee, and ankle joints, and generate the power needed for movement.

1. Sartorius

2. Iliopsoas

3. Pectineus

4. Adductor longus

5. Adductor magnus

6. Gracialis

7. Rectus femoris

8. Vastus lateralis

9. Vastus medialis

10. Peroneus longus

11. Tibialis anterior

12. Gastrocnemius

13. Extensor digitorum brevis

14. Extensor hallucis brevis

15. Gluteus maximus

16. Biceps femoris

17. Plantaris

18. Soleus

Interesting Fact

The gluteus maximus, the most significant muscle in the body, is primarily responsible for extending the thigh. It plays a crucial role in actions such as standing up from a seated position, climbing stairs, and running.

Muscles of the Lower Limb

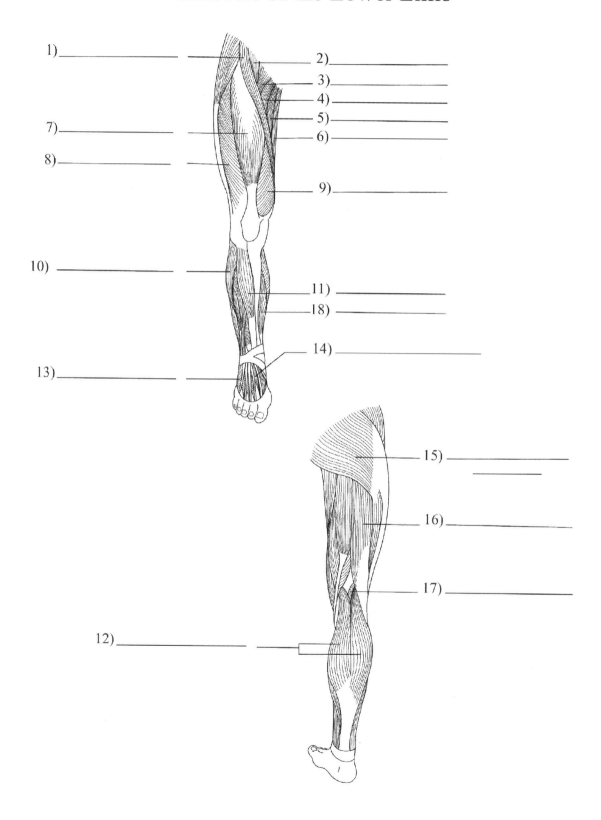

1) _____

2) _____

3) _____

4) _____

5) _____

6) _____

7) _____

8) _____

9) _____

10) _____

11) _____

18) _____

14) _____

13) _____

15) _____
_____

16) _____

17) _____

12) _____

Muscles of the Foot

The foot muscles are crucial for supporting the foot's arches, maintaining balance, and facilitating movements such as walking, running, and jumping. These muscles coordinate with the foot's bones, ligaments, and tendons to provide stability and flexibility during various activities.

1. Tibialis anterior

2. Extensor digitorum longus

3. Extensor hallucis longus

4. Extensor digitorum brevis

5. Fibularis longus

6. Fibularis brevis

7. Abductor hallucis

8. Flexor digitorum brevis

9. Quadratus plantae

10. Lumbricals

11. Flexor hallucis brevis

12. Flexor digiti minimi brevis

Interesting Fact

The foot's intrinsic muscles, such as the flexor hallucis brevis, play a vital role in maintaining the foot's arches and providing support during locomotion. These small yet powerful muscles contribute to the foot's overall strength and agility.

Muscles of the foot

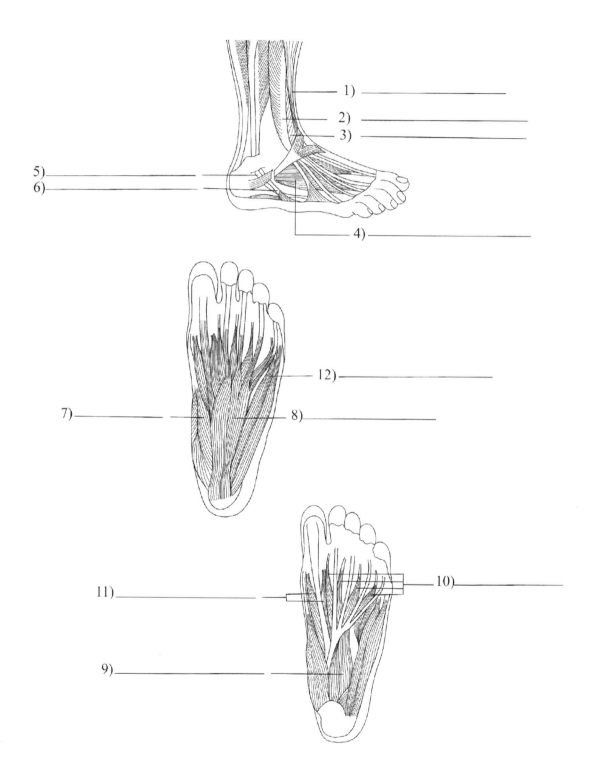

1)

2)

3)

5)

6)

4)

12)

7)

8)

11)

10)

9)

Muscles of the Pelvis

The pelvic muscles play a critical role in supporting the pelvic organs, maintaining posture, and stabilizing the hip joints. These muscles are essential for various movements, including walking, sitting, and standing, and they also contribute to the overall stability and function of the pelvis.

1. Iliacus

2. Psoas minor

3. Psoas major

4. Piriformis

5. Obturator internus

6. Sartorius

7. Coccygeus

8. Iliococcygeus

9. Pubococcygeus

10. Puburectalis

11. Adductor longus

12. Adductor brevis

Interesting Fact

The piriformis muscle, located deep within the buttocks, is named after its pear-like shape. Interestingly, the piriformis muscle is also involved in a condition known as piriformis syndrome, which can cause pain and discomfort in the buttocks and lower back.

Muscles of the pelvis

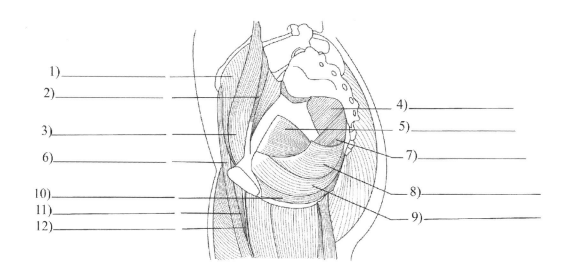

1)

2)

3)

6)

10)

11)

12)

4)

5)

7)

8)

9)

The Heart

The heart is a vital organ that pumps blood throughout the body and delivers oxygen and nutrients to tissues and organs while removing waste products. Understanding the heart's anatomy is crucial for comprehending its function in the circulatory system.

1. Right atrium

2. Right ventricle

3. Left atrium

4. Left ventricle

5. Tricuspid valve

6. Mitral valve

7. Pulmonary valve

8. Aorta

9. Left pulmonary artery

10. Left pulmonary vein

11. Right pulmonary artery

12. Right pulmonary vein

13. Superior Vena Cava

14. Interventricular septum

15. Chordae tendinae

16. Aortic Valve

17. Inferior Vena Cava

18. Descending aorta

Interesting Fact

The heart is approximately the size of a fist and beats about 100,000 times daily, pumping around 7,500 liters (2,000 gallons) of blood through the body daily.

The heart

13) _____

11) _____

7) _____

12) _____

1) _____

5) _____

15) _____

2) _____

8) _____

9) _____

10) _____

3) _____

6) _____

4) _____

16) _____

14) _____

18) _____

17) _____

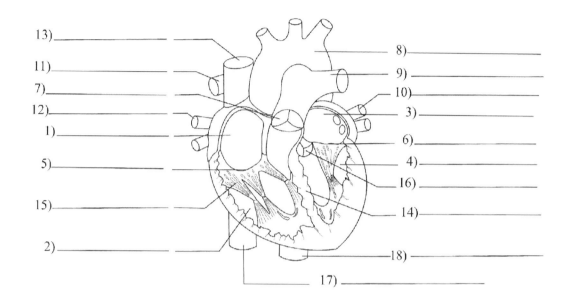

5) _____

16) _____

6) _____

7) _____

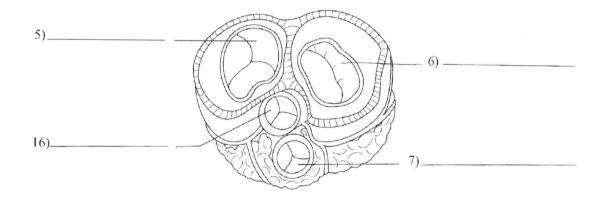

The Coronary Blood Vessels

The coronary blood vessels supply the heart muscle (myocardium) with oxygenated blood, ensuring its proper function and vitality. Understanding the anatomy of these vessels is essential for comprehending blood circulation within the heart and diagnosing and treating cardiovascular diseases.

1. Right Coronary Artery
2. Conus Arteriosus branch
3. Sinoatrial nodal branch
4. Right Atrial Branches
5. Right Ventricular Branches
6. Right Marginal Branches
7. Posterior Interventricular Artery
8. Atrioventricular Nodal Branch
9. Right Posterolateral Branch
10. Left coronary artery
11. Circumflex artery
12. Left marginal branch
13. Diagonal branch
14. Left anterior descending artery
15. Interventricular septal branches
16. Posterior descending artery
17. Left posterior ventricular branch
18. Right posterolateral branch

Interesting Fact

The coronary arteries have a unique ability to dilate or constrict in response to changes in demand, ensuring that the heart receives adequate blood flow during rest and activity.

The coronarty blood vesles

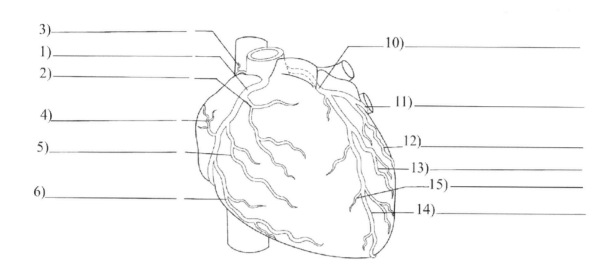

3)
1)
2)
4)
5)
6)

10)
11)
12)
13)
15)
14)

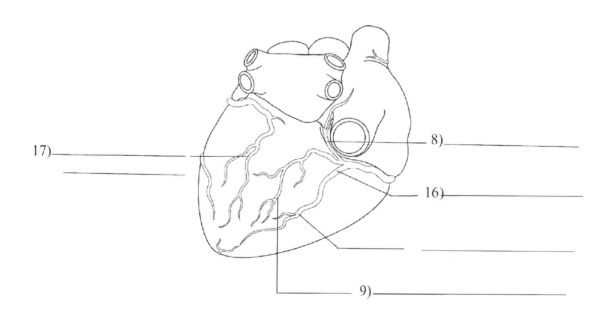

17)

8)
16)

9)

Blood vessels of the Head and Neck

The arteries and veins of the head and neck form a complex network that ensures blood supply to the brain, facial structures, and surrounding tissues.

1. Facial artery

2. Maxillary artery

3. Superficial temporal artery

4. Occipital artery

5. Lingual artery

6. Supraorbital artery

7. Posterior auricular artery

8. Ascending pharyngeal artery

9. External carotid artery

10. Internal carotid artery

11. Superior thyroid artery

12. Vertebral artery

13. Superficial temporal vein

14. Occipital vein

15. Facial vein

16. Angular vein

17. Superior thyroid vein

18. Internal jugular vein

19. External jugular vein

20. Retromandibular vein

Interesting Fact

The superficial temporal artery, a branch of the external carotid artery, is commonly used as a landmark for pulse palpation during physical examination.

Blood vessels off the head and neck

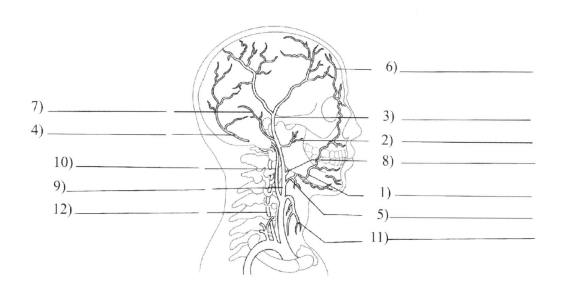

7) _____
4) _____
10) _____
9) _____
12) _____

6) _____
3) _____
2) _____
8) _____
1) _____
5) _____
11) _____

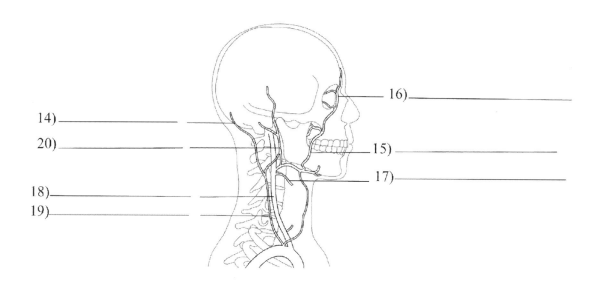

14) _____
20) _____
18) _____
19) _____

16) _____
15) _____
17) _____

The Arteries of Upper Limb

The arteries of the upper limb supply oxygenated blood to the muscles, bones, and tissues of the arm, forearm, wrist, and hand. Let's explore them.

1. Subclavian artery

2. Axillary artery

3. Brachial artery

4. Deep brachial artery

5. Humeral Circumflex artery

6. Radial artery

7. Anterior ulnar recurrent artery

8. Posterior ulnar recurrent artery

9. Ulnar artery

10. Anterior crural interosseus

11. Deep palmar arch

12. Superior palmar arch

13. Princeps Pollicis artery

14. Radialis indicis artery

15. Common digital artery

16. Proper digital artery

Interesting Fact

The brachial artery, a major upper limb artery, is commonly used for measuring blood pressure. It can be palpated in the antecubital fossa, making it easily accessible for healthcare professionals during routine examinations.

Arteries of the upper limb

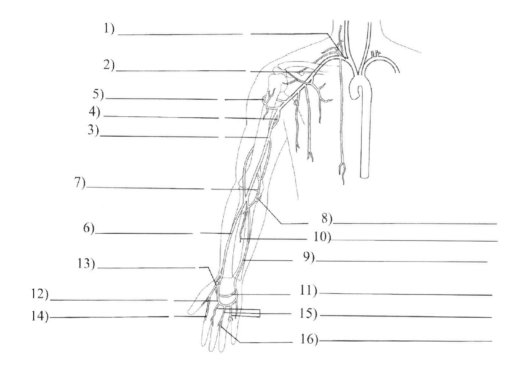

1) _____ _____

2) _____ _____

5) _____

4) _____

3) _____

7) _____

6) _____

13) _____

12) _____

14) _____

8) _____

10) _____

9) _____

11) _____

15) _____

16) _____

The Arteries of Lower Limb

The arteries of the lower limb supply oxygenated blood to the muscles, bones, and tissues of the thigh, leg, ankle, and foot. Let's go into the details.

1. Common iliac artery

2. External iliac artery

3. Internal iliac artery

4. Femoral artery

5. Deep femoral artery

6. Lateral circumflex artery

7. Medial circumflex artery

8. Genicular artery

9. Popliteal artery

10. Anterior tibial artery

11. Posterior tibial artery

12. Fibular artery

13. Dorsalis pedis artery

14. Dorsal arch

15. Plantar arch

16. Peroneal artery

17. Medial plantar artery

18. Lateral plantar artery

Interesting Fact

The femoral artery, one of the major arteries of the lower limb, is located in the femoral triangle, making it accessible for arterial access. It is often used as a site for arterial catheterization during medical procedures such as cardiac catheterization and angioplasty.

The arteries of lower limb

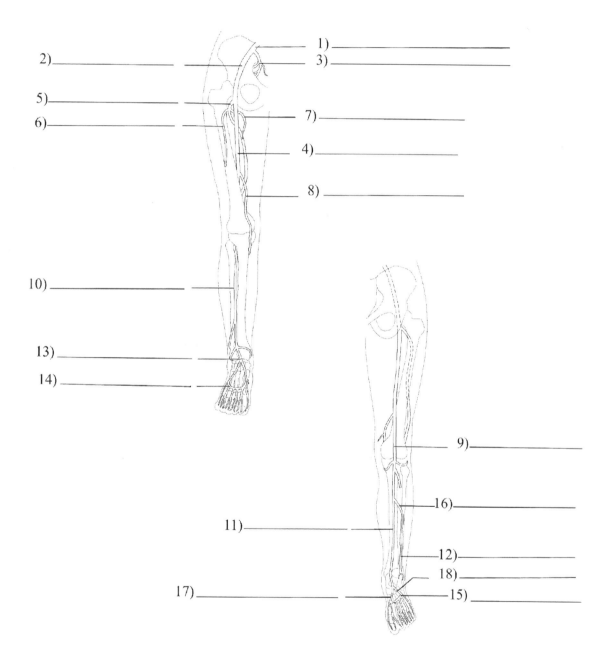

2)_____

5)_____

6)_____

10)_____

13)_____

14)_____

1)_____

3)_____

7)_____

4)_____

8)_____

9)_____

16)_____

11)_____

12)_____

18)_____

17)_____

15)_____

Venous drainage of the Upper Limb

The venous drainage of the upper limb returns deoxygenated blood from the arm, forearm, wrist, and hand back to the heart. Here are the common veins present in the upper limb.

1. Subclavian vein

2. Axillary vein

3. Cephalic vein

4. Subscapular vein

5. Brachial vein

6. Basilic vein

7. Median cubital vein

8. Radial vein

9. Medial antebrachial vein

10. Ulnar vein

11. Superior palmar venous arch

12. Deep palmar venous arch

13. Digital vein

Interesting Fact

The basilic vein, one of the major veins of the upper limb, is often used for venous access during medical procedures such as blood draws and intravenous therapy. It runs along the medial aspect of the arm and is relatively large and superficial, making it suitable for venipuncture.

Venous drainage of the upper limb

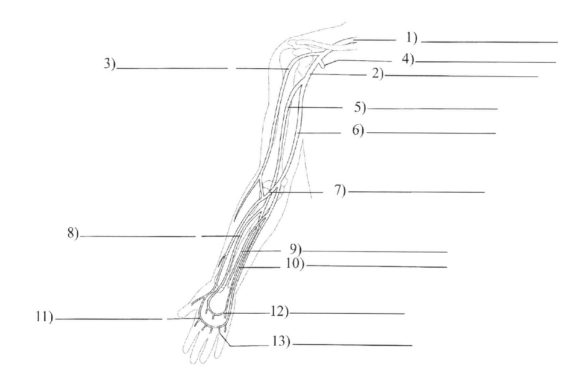

1) _____

4) _____

3) _____

2) _____

5) _____

6) _____

7) _____

8) _____

9) _____

10) _____

11) _____

12) _____

13) _____

Venous drainage of the Lower Limb

The venous drainage of the lower limb returns deoxygenated blood from the thigh, leg, ankle, and foot back to the heart. Here are the common veins present in the lower limb.

1. Common iliac vein

2. Internal iliac vein

3. External iliac vein

4. Great saphenous vein

5. Small saphenous vein

6. Deep femoral vein

7. Femoral circumflex vein

8. Femoral vein

9. Popliteal vein

10. Gluteal vein

11. Anterior tibial vein

12. Posterior tibial vein

13. Fibular vein

14. Deep venous arch of the foot

15. Dorsal venous arch of the foot

16. Lateral plantar vein

17. Medial plantar vein

18. Plantar venous arches

Interesting Fact

The great saphenous vein, the longest vein in the body, travels along the medial aspect of the lower limb and is often used for surgical procedures such as coronary artery bypass grafting (CABG) and venous grafting.

Venous drainage of the lower limb

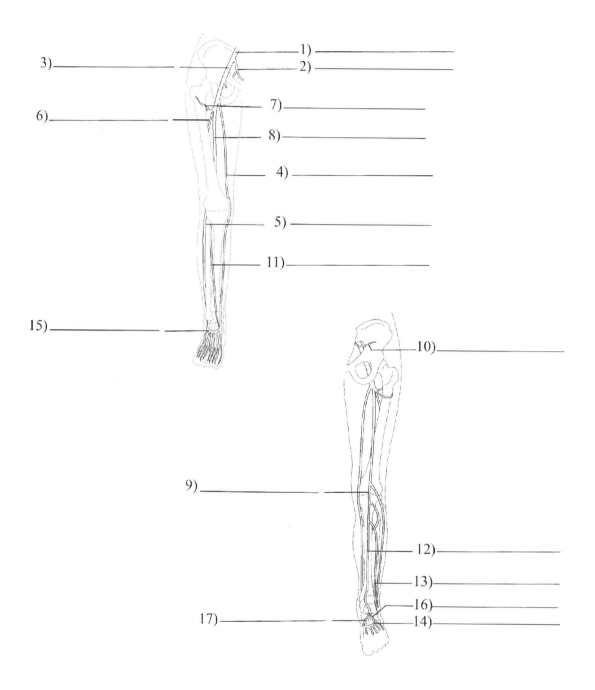

3) _____

6) _____

15) _____

1) _____

2) _____

7) _____

8) _____

4) _____

5) _____

11) _____

10) _____

9) _____

12) _____

13) _____

16) _____

17) _____

14) _____

The Lymphatic System

The lymphatic system plays a crucial role in maintaining fluid balance, immune function, and waste removal in the body. Comprising lymphatic vessels, lymph nodes, and lymphoid organs, this system serves as a network for transporting lymphatic fluid and immune cells throughout the body.

1. Spleen

2. Thymus

3. Thoracic duct

4. Pharyngeal tonsil

5. Lingual tonsil

6. Palatine tonsil

7. Cervical lymph node

8. Axillary lymph node

9. Intestinal lymph node

10. Vermiform appendix

11. Bone marrow

12. Peyer's patches

13. Cisterna chyli

14. Inguinal lymph node

15. Lymphatic vessel

Interesting Fact

The thoracic duct, the largest lymphatic vessel in the body, drains lymph from the lower body, left upper body, and left side of the head and neck into the venous circulation. It begins in the abdomen and ascends through the thorax to empty into the left subclavian vein.

The lymphatic system

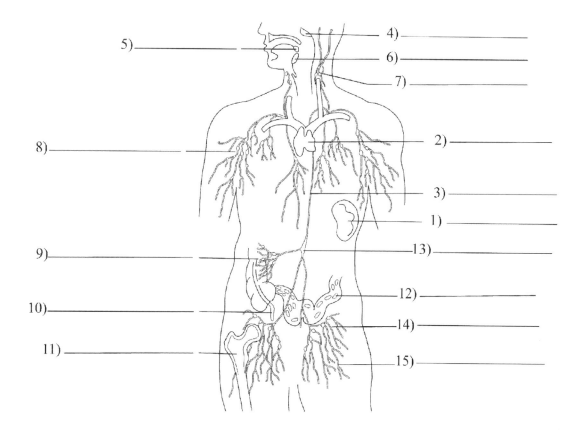

5) ————————————

4) ————————————————————

6) ————————————————————

7) ————————————————————

8) ————————————————————

2) ————————————————————

3) ————————————————————

1) ————————————————————

9) ————————————

13) ————————————————

10) ————————————

12) ————————————————

14) ————————————————

11) ————————————

15) ————————————————

Upper Respiratory Tract

The upper respiratory tract comprises the organs and structures responsible for the initial respiration processes, including air intake, filtration, and humidification. It serves as the gateway for air to enter the respiratory system and is crucial for maintaining proper respiratory function.

1. Nasal cavity

2. Olfactory epithelium

3. Superior Nasal Concha

4. Middle Nasal Concha

5. Inferior Nasal Concha

6. Hard palate

7. Soft palate

8. Oral cavity

9. Epiglottis

10. Glottis

11. Nasopharynx

12. Oropharynx

13. Laryngopharynx

14. Frontal sinus

15. Sphenoid sinus

16. Trachea

17. Hyoid bone

Interesting Fact

The nasal cavity, lined with mucous membranes and tiny hair-like structures called cilia, plays a vital role in filtering and humidifying the air we breathe. It also contains olfactory receptors, enabling us to detect and distinguish various scents.

Upper respiratory tract

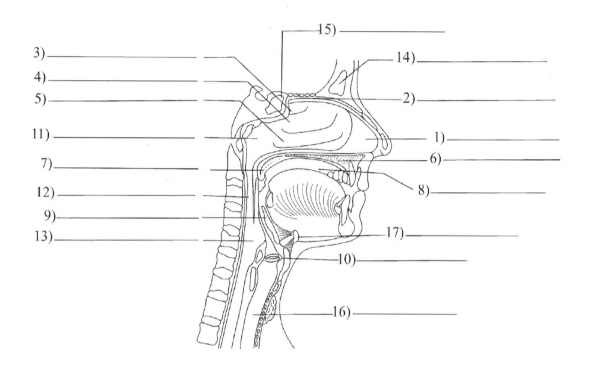

3)

4)

5)

11)

7)

12)

9)

13)

15)

14)

2)

1)

6)

8)

17)

10)

16)

Lower Respiratory Tract

The upper respiratory tract comprises the organs and structures responsible for the initial respiration processes, including air intake, filtration, and humidification. It serves as the gateway for air to enter the respiratory system and is crucial for maintaining proper respiratory function. Let's explore the parts.

1. Trachea

2. Thyroid cartilage

3. Cricoid cartilage

4. Primary bronchi

5. Secondary bronchi

6. Bronchioles

7. Right upper lobe

8. Right middle lobe

9. Right lower lobe

10. Left upper lobe

11. Left lower lobe

12. Alveoli

13. Alveolar duct

14. Capillaries

15. Diaphragm

Interesting Fact

An average adult human lung contains approximately 300 million alveoli, providing an extensive area for gas exchange.

Lower respiratory tract

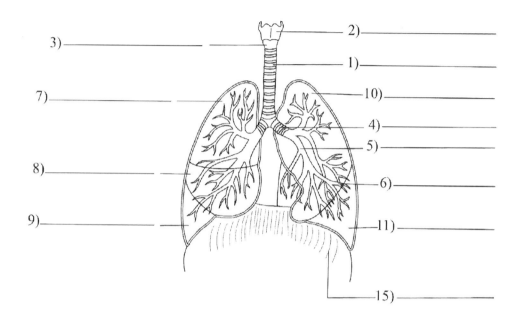

3)———————————
2)———————————
1)———————————
7)———————————
10)———————————
4)———————————
5)———————————
8)———————————
6)———————————
9)———————————
11)———————————
15)———————————

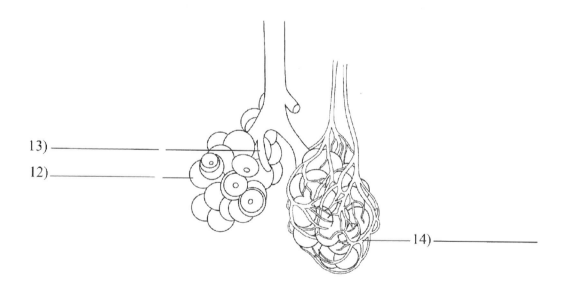

13)———————————
12)———————————
14)———————————

Oral Cavity

The oral cavity, also known as the mouth, is a complex structure involved in various functions such as ingestion, chewing, speech, and digestion.

1. Incisor

2. Canine

3. Premolar

4. Molar

5. Tongue

6. Gingiva

7. Lingual frenum

8. Inferior labial frenum

9. Superior labial frenum

10. Hard palate

11. Soft palate

12. Uvula

13. Oropharynx

14. Palatine tonsil

15. Palatopharyngeal arch

16. Vestibule

17. Parotid gland

18. Submandibular gland

19. Sublingual gland

Interesting Fact

The tongue is not only crucial for tasting food but also plays a vital role in speech articulation and swallowing. It contains numerous taste buds that detect sweet, sour, salty, and bitter flavors, contributing to the sensory experience of eating.

Oral cavity

9)
10)
11)
15)
5)
7)
2)
16)

6)
14)
12)
13)
4)
3)
1)
8)

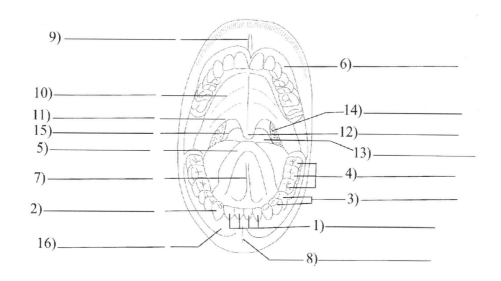

17)
19)
18)

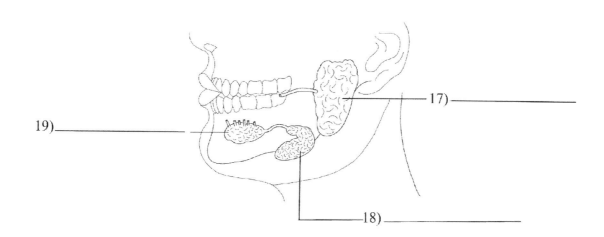

Pharynx and Esophagus

The pharynx and esophagus are important components of the digestive and respiratory systems. They play key roles in swallowing and passing food and liquids from the mouth to the stomach.

1. Nasopharynx

2. Oropharynx

3. Laryngopharynx

4. Hyoid bone

5. Larynx

6. Esophagus

7. Trachea

8. Epiglottis

9. Palatine tonsil

10. Pharyngeal tonsil

Interesting Fact

The epiglottis is a flap-like structure located at the base of the tongue that prevents food and liquids from entering the airway (trachea) during swallowing. It closes off the entrance to the trachea, directing substances toward the esophagus for safe passage into the stomach.

Pharynx and esophagus

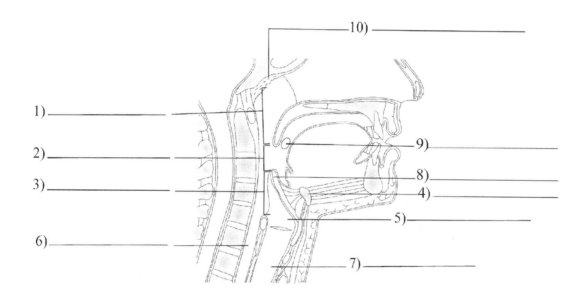

1)

2)

3)

6)

10)

9)

8)

4)

5)

7)

Stomach

The stomach is a muscular organ in the upper abdomen. It is the primary site for food storage, digestion, and mixing with gastric juices. Let's explore the anatomy below.

1. Cardia

2. Fundus

3. Antrum

4. Pylorus

5. Body

6. Longitudinal muscle layer

7. Circular muscle layer

8. Oblique muscle layer

9. Pyloric sphincter

10. Rugae

11. Gastric pit

12. Gastric gland

13. Connective tissue layer

14. Parietal cell

15. Chief cell

16. Enteroendocrine cell

Interesting Fact

A layer of mucus secreted by specialized cells protects the stomach lining from the acidic environment. However, excessive alcohol consumption or certain medications can lead to the erosion of this protective layer, resulting in conditions such as gastritis or peptic ulcers.

Stomach

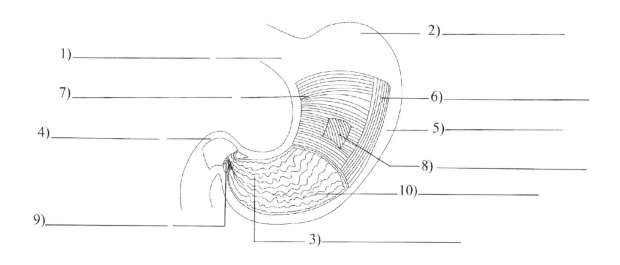

1) _____

2) _____

7) _____

6) _____

4) _____

5) _____

8) _____

10) _____

9) _____

3) _____

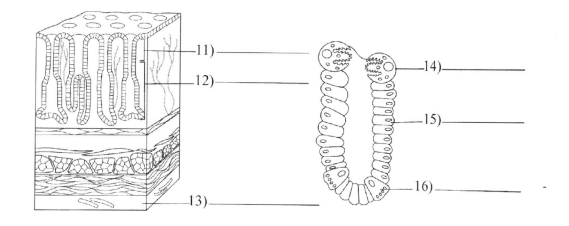

11) _____

14) _____

12) _____

15) _____

13) _____

16) _____

Intestine

The small and large intestines, play vital roles in the digestion and absorption of nutrients and elimination of waste products from the body. Let's explore the anatomy below.

1. Duodenum

2. Jejunum

3. Ileum

4. Transverse colon

5. Descending colon

6. Ascending colon

7. Sigmoid colon

8. Cecum

9. Appendix

10. Rectum

11. Serosa

12. Longitudinal muscle

13. Circular muscle

14. Mucosa

15. Submucosa

16. Villi

17. Lacteal

18. Epithelial cell

19. Capillary network

20. Lymphatic vessel

Interesting Fact

The small intestine is the digestive system's primary site for nutrient absorption. Despite its relatively small diameter, its extensive surface area, thanks to villi and microvilli, allows for efficient absorption of nutrients into the bloodstream.

Intestine

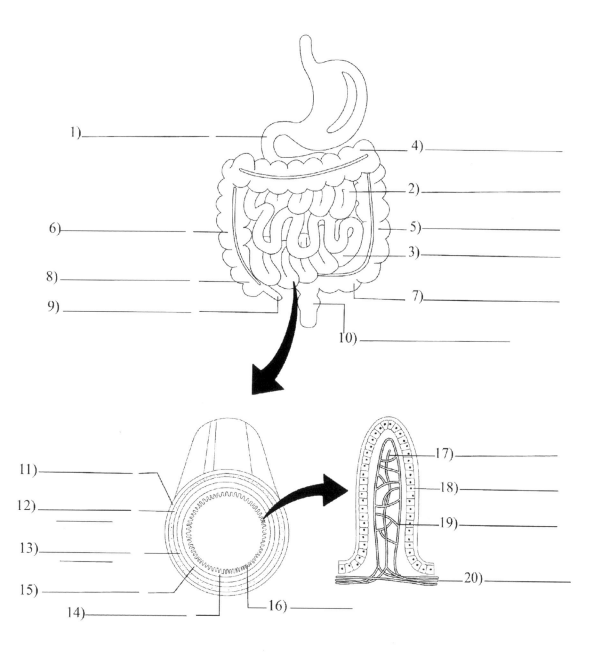

1)

4)

2)

6)

5)

3)

8)

7)

9)

10)

11)

17)

12)

18)

13)

19)

15)

20)

14)

16)

Accessory Digestive Organs

The accessory digestive organs, including the liver, gallbladder, and pancreas, play essential roles in the digestion and metabolism of nutrients and regulating digestive processes. Let's explore the anatomy below.

1. Liver

2. Gall bladder

3. Pancreas

4. Common hepatic duct

5. Spleen

6. Pancreatic duct

7. Cystic duct

8. Common bile duct

9. Duodenum

10. Diaphragm

Interesting Fact

The liver is the largest internal organ in the body and performs numerous vital functions, including detoxification, nutrient metabolism, and bile production.

Accessory digestive organs

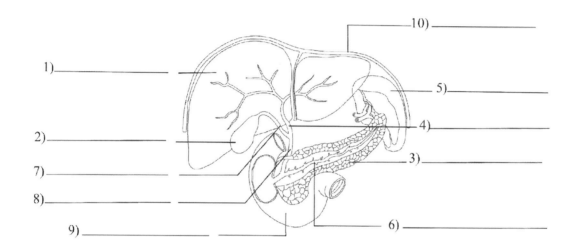

1) _____

2) _____

7) _____

8) _____

9) _____

10) _____

5) _____

4) _____

3) _____

6) _____

Liver

The liver is a vital organ located in the upper right quadrant of the abdomen. It plays a central role in metabolism, detoxification, and digestion. Let's explore the anatomy below.

1. Right lobe

2. Left lobe

3. Caudate lobe

4. Falciform ligament

5. Teres ligament

6. Inferior vena cava

7. Aorta

8. Gall bladder

9. Portal vein

10. Quadrate lobe

11. Common bile duct

12. Proper hepatic artery

Interesting Fact

The liver has remarkable regenerative capacity. It can regenerate lost tissue and regain its functional capacity even after significant damage or surgical removal. This unique ability makes liver transplantation and recovery possible.

Liver

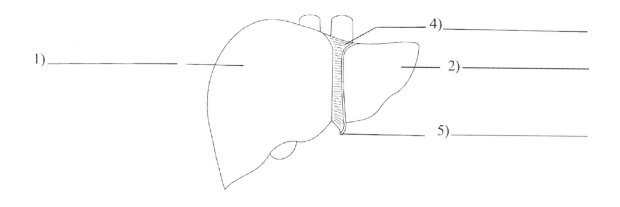

1) _____

4) _____

2) _____

5) _____

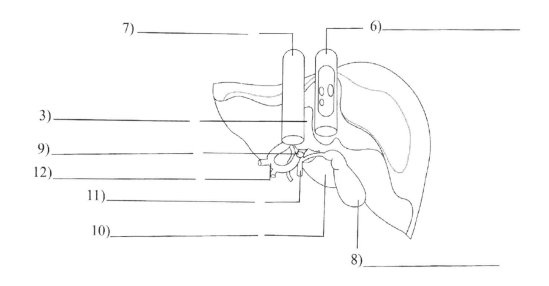

7) _____

6) _____

3) _____

9) _____

12) _____

11) _____

10) _____

8) _____

The Renal System

The renal system, comprising the kidneys, ureters, bladder, and urethra, plays a vital role in maintaining fluid and electrolyte balance, regulating blood pressure, and removing metabolic waste from the body. Let's explore the anatomy below.

1. Left kidney

2. Right kidney

3. Renal artery

4. Renal vein

5. Urethra

6. Urinary bladder

7. Inferior vena cava

8. Aorta

9. Left Adrenal gland

10. Right adrenal gland

11. Renal cortex

12. Renal medulla

13. Renal pelvis

14. Hepatic vein

Interesting Fact

Each kidney contains millions of functional units called nephrons, which are responsible for filtering blood and producing urine.

The renal system

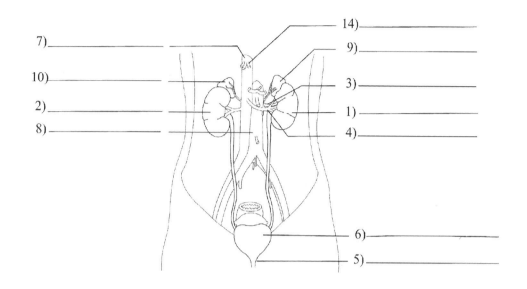

7) _____
10) _____
2) _____
8) _____

14) _____
9) _____
3) _____
1) _____
4) _____

6) _____
5) _____

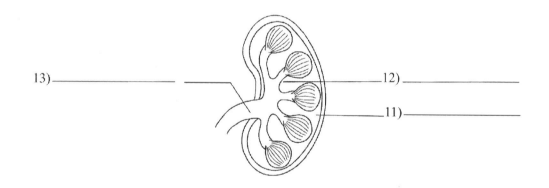

13) _____

12) _____
11) _____

The Nephron

The nephron is the functional unit of the kidney, responsible for filtering blood and producing urine. It is crucial in regulating the body's fluid balance, electrolyte levels, and pH. Let's explore the anatomy below.

1. Renal corpuscle

2. Glomerulus

3. Bowman's capsule

4. Proximal convoluted tubule

5. Loop of Henle

6. Distal convoluted tubule

7. Collecting duct

8. Afferent arteriole

9. Efferent arteriole

10. Vasa recta

11. Peritubular capillaries

12. Interlobular vein

13. Cortex

14. Medulla

15. Renal pyramid

Interesting Fact

Each kidney contains approximately one million nephrons, highlighting the remarkable efficiency of the renal system in maintaining homeostasis.

The nephron

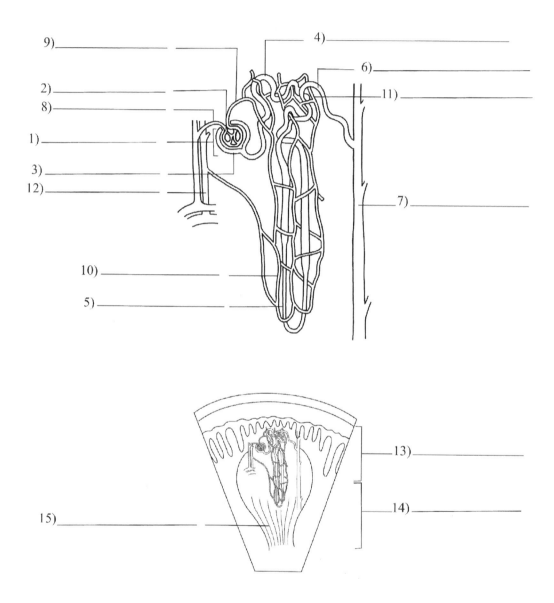

9) _____

4) _____

6) _____

2) _____

8) _____

11) _____

1) _____

3) _____

12) _____

7) _____

10) _____

5) _____

13) _____

14) _____

15) _____

The Female Reproductive System

The female reproductive system is a complex network of organs responsible for producing eggs, fertilization, gestation, and childbirth. It comprises several structures working in harmony to support the reproductive process. Let's explore the anatomy below.

1. Ovary

2. Fallopian tube

3. Uterus

4. Cervix

5. Cervical canal

6. Vagina

7. Clitoris

8. Endometrium

9. Labium Minora

10. Labium Majora

11. Urethral opening

12. Vaginal opening

Interesting Fact

The ovaries, a vital component of the female reproductive system, are responsible for producing eggs and secreting hormones like estrogen and progesterone. These hormones regulate the menstrual cycle and influence various aspects of female physiology.

Female reproductive system

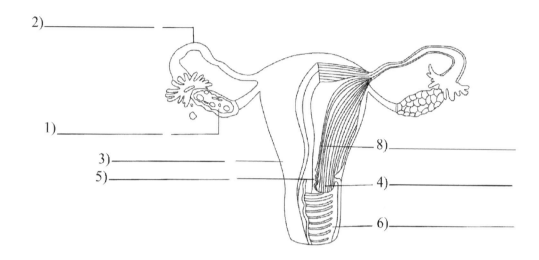

2) _____

1) _____

3) _____

5) _____

8) _____

4) _____

6) _____

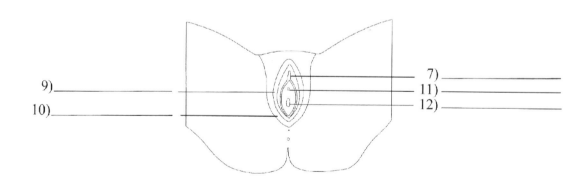

9) _____

10) _____

7) _____

11) _____

12) _____

The Male Reproductive System

The male reproductive system consists of organs that work together to produce and deliver sperm, allowing for the fertilization of a female egg. It is a vital component of human reproduction and contributes to genetic diversity. Let's explore the anatomy below.

1. Testes
2. Epididymis
3. Vas deferens
4. Seminal vesicles
5. Prostate gland
6. Urethra
7. Bulbourethral glands
8. Penis
9. Glans penis
10. Scrotum
11. Epididymis
12. Ductus deferens
13. Seminal vesicle
14. Ejaculatory duct
15. Urinary bladder
16. Tunica vaginalis
17. Efferent ductule
18. Tunica Albuginea
19. Seminiferous tubule
20. Rete testes

Interesting Fact

The testes, the primary male reproductive organs, produce sperm and secrete testosterone, the hormone responsible for male secondary sexual characteristics such as facial hair growth and deepening of the voice.

Male reproductive system

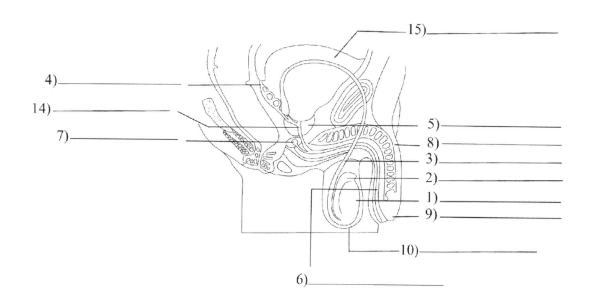

4)_____

14)_____

7)_____

15)_____

5)_____

8)_____

3)_____

2)_____

1)_____

9)_____

10)_____

6)_____

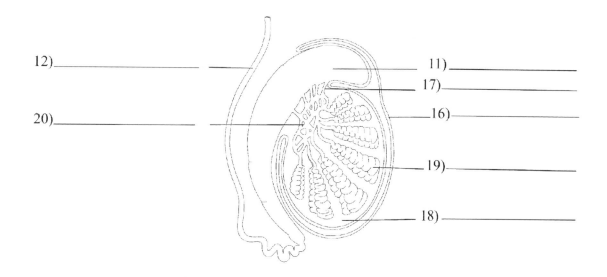

12)_____

20)_____

11)_____

17)_____

16)_____

19)_____

18)_____

The Ovary and Breast

The ovaries and breasts are essential components of the female reproductive system, each with unique reproduction and hormonal regulation functions. Let's explore the anatomy below.

1. Primordial follicle

2. Primary follicle

3. Developing follicle

4. Secondary ovum

5. Graafian follicle

6. Ruptured follicle

7. Early corpus luteum

8. Corpus luteum

9. Corpus albicans

10. Fat tissue

11. Alveoli

12. Lactiferous ducts

13. Lobule

14. Lactiferous sinus

15. Areola

Interesting Fact

Both the ovary and breast undergo significant changes throughout a woman's life. The ovaries experience monthly hormonal fluctuations during the menstrual cycle, while the breasts undergo.

Ovary and breasts

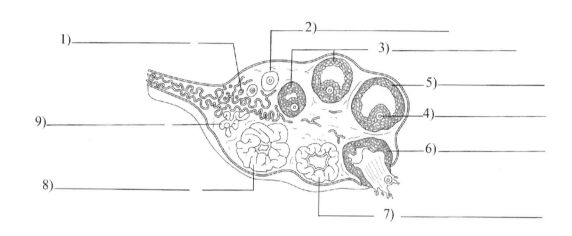

1) _____
2) _____
3) _____
5) _____
4) _____
6) _____
9) _____
8) _____
7) _____

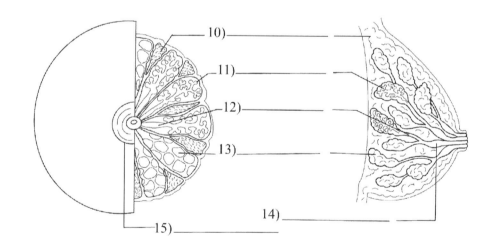

10) _____
11) _____
12) _____
13) _____
14) _____
15) _____

The Endocrine System

The endocrine system is a complex network of glands and organs responsible for secreting hormones that regulate various physiological processes in the body. These hormones act as chemical messengers, coordinating activities such as metabolism, growth and development, reproduction, and mood regulation. Let's explore the anatomy below.

1. Hypothalamus

2. Pituitary gland

3. Pineal gland

4. Thyroid gland

5. Parathyroid gland

6. Adrenal gland

7. Pancreas

8. Thymus gland

9. Ovary

10. Testis

11. Optic chiasm

12. Anterior pituitary

13. Posterior pituitary

14. Sella turcica

Interesting Fact

The pituitary gland, often called the "master gland," plays a crucial role in regulating other endocrine glands. It secretes several hormones that influence growth, reproduction, and metabolism.

Endocrine system

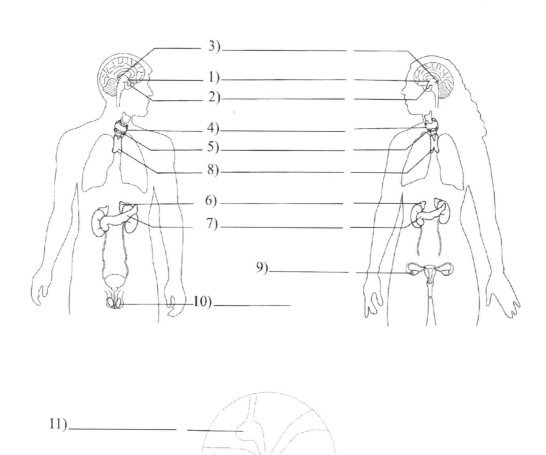

3)
1)
2)
4)
5)
8)
6)
7)
9)
10)

11)
12)
13)
14)

Exocrine Glands

Exocrine glands are vital to the body's physiology. They secrete substances onto epithelial surfaces or into ducts that lead to body cavities or the external environment. These glands play diverse roles in digestion, lubrication, and protection of various tissues and organs.

1. Simple tubular

2. Simple branched tubular

3. Simple coiled tubular

4. Simple acinar

5. Simple branched acinar

6. Compound tubular

7. Compound acinar

8. Compound tubuloacinar

Interesting Fact

The salivary glands, a type of exocrine gland, produce saliva containing enzymes that aid in the digestion of carbohydrates, highlighting the importance of exocrine glands in maintaining digestive health.

Exocrine glands

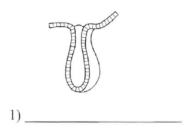

1) _____

2) _____

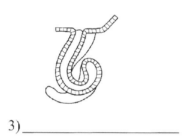

3) _____

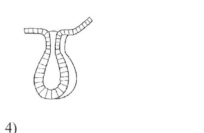

4) _____

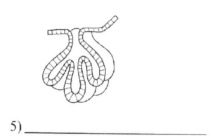

5) _____

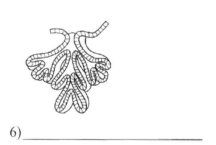

6) _____

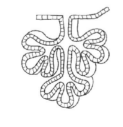

7) _____

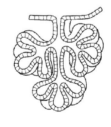

8) _____

The Integumentary System

The integumentary system is the body's most extensive organ system, encompassing the skin and its appendages, including hair, nails, sweat glands, and sebaceous glands. It serves as a protective barrier against external threats, regulates body temperature, and plays a role in sensory perception and vitamin D synthesis.

1. Epidermis

2. Dermis

3. Hypodermis

4. Hair follicle

5. Hair root

6. Hair follicle receptor

7. Hair shaft

8. Eccrine sweat gland

9. Sebaceous gland

10. Pacinian corpuscle

11. Cutaneous vascular plexus

12. Arrector pili muscle

13. Adipose tissue

14. Sensory nerve fiber

15. Nail bed

16. Nail plate

17. Lunula

18. Cuticle

19. Nail root

20. Nail matrix

Interesting Fact

The skin is a dynamic organ capable of self-renewal and repair. It constantly sheds dead skin cells and regenerates new ones, demonstrating its remarkable ability to adapt and maintain homeostasis.

Integumentary system

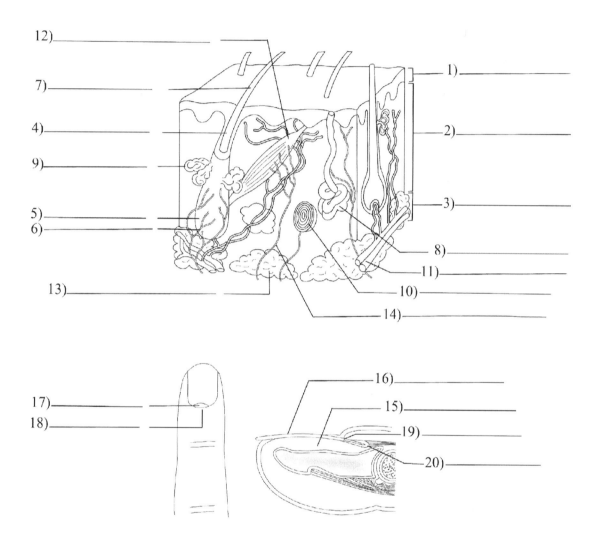

The Nervous System

The nervous system is a complex network of cells and tissues that coordinates and controls the body's actions and responses to external stimuli. It comprises the central nervous system (brain and spinal cord) and the peripheral nervous system (nerves and ganglia), working together to process information and transmit signals throughout the body.

1. Brain

2. Spinal cord

3. Ganglion

4. Nerve

5. Nerve cell or Neuron

6. Dendrites

7. Cell body or soma

8. Nucleus

9. Axon

10. Myelin sheath

11. Axon terminal

Interesting Fact

Neurons, the building blocks of the nervous system, are highly specialized cells capable of transmitting electrical and chemical signals. Despite their diverse functions, neurons share standard structural features, such as dendrites, cell bodies, and axons, that enable communication within the nervous system.

Nervous system

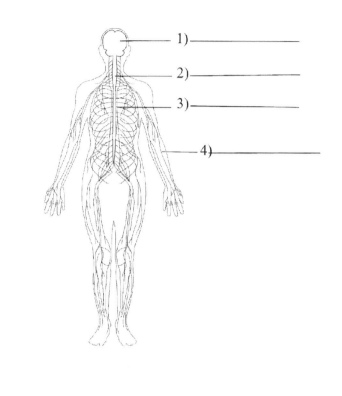

1) _____

2) _____

3) _____

4) _____

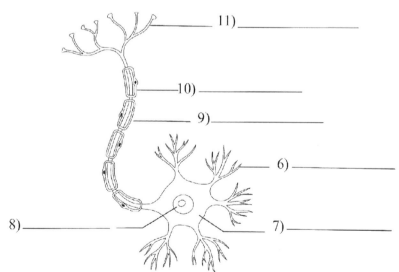

11) _____

10) _____

9) _____

6) _____

8) _____

7) _____

The Brain

The brain is the central organ of the nervous system, responsible for processing information, controlling bodily functions, and coordinating behavior. It comprises different regions, each with specialized functions, working together to support cognition, emotion, and motor skills. Let's explore its anatomy.

1. Cerebrum

2. Cerebellum

3. Temporal lobe

4. Brain stem

5. Midbrain

6. Pons

7. Medulla oblongata

8. Frontal lobe

9. Parietal lobe

10. Temporal lobe

11. Occipital lobe

12. Corpus callosum

13. Optic chiasm

14. Limbic lobe

15. Precentral gyrus

16. Post central gyrus

17. Arachnoid mater

18. Dura mater

19. Pia mater

20. Subarachnoid space

21. Longitudinal fissure

22. Cerebral vein

Interesting Fact

The human brain is one of the most complex structures in the known universe, containing billions of neurons and trillions of synaptic connections.

Brain

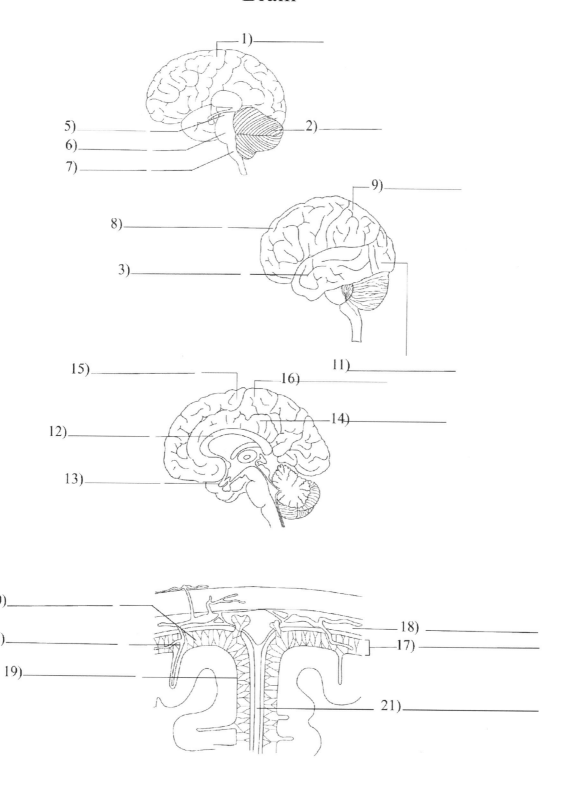

The Spinal Cord

The spinal cord is a vital component of the central nervous system, serving as a conduit for transmitting signals between the brain and the rest of the body. It plays a crucial role in sensory perception, motor control, and reflex responses to stimuli.

1. Gray matter

2. White matter

3. Dorsal nerve root

4. Dorsal nerve root ganglion

5. Ventral nerve root

6. Dorsal ramus

7. Ventral ramus

8. Spinal nerve

9. Central canal

10. Lateral horn

11. Anterior horn

Interesting Fact

The spinal cord is protected by the vertebral column (spine), consisting of bony vertebrae stacked on each other. This structure provides both support and flexibility while safeguarding the delicate nerves of the spinal cord.

Spinal cord

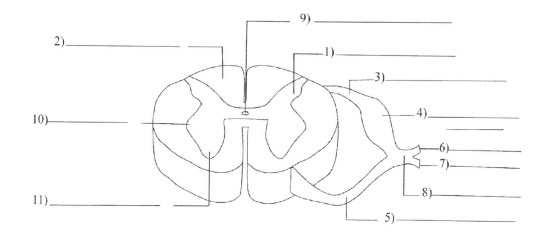

The Eye

The eye is a complex sensory organ responsible for vision. It allows us to perceive light and distinguish shapes, colors, and depth. It comprises several specialized structures that work together to capture, focus, and transmit visual information to the brain for interpretation.

1. Cornea

2. Iris

3. Pupil

4. Lens

5. Anterior chamber

6. Posterior chamber

7. Suspensory ligament

8. Conjunctiva

9. Sclera

10. Ciliary body

11. Choroid

12. Lateral rectus muscle

13. Retina

14. Fovea centralis

15. Optic nerve

16. Optic disc or blind spot

17. Medial rectus muscle

18. Rod

19. Cone

20. Ganglion cell

21. Bipolar cell

22. Retinal pigment epithelium

Interesting Fact

The human eye is often compared to a camera due to its remarkable ability to focus light onto the retina, where photoreceptor cells convert it into electrical signals. This process enables us to perceive the world with incredible detail and clarity.

Eye

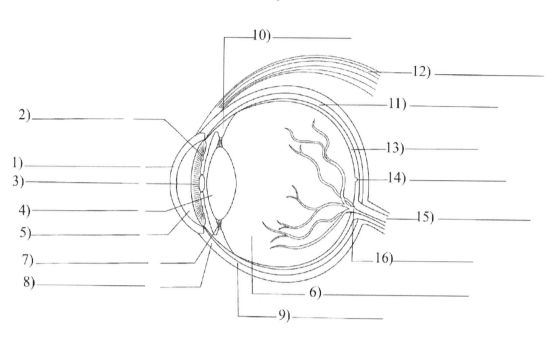

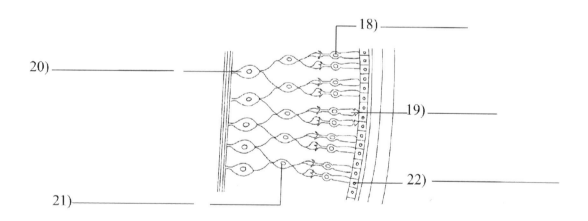

The Ear

The ear is a complex sensory organ responsible for hearing and balance. It allows us to perceive sound and maintain equilibrium. It comprises three main regions: the outer ear, middle ear, and inner ear, each with distinct structures that contribute to auditory and vestibular function.

1. Auricle

2. External acoustic meatus

3. Elastic cartilage

4. Auditor ossicles

5. Semi-circular canals

6. Vestibule

7. Vestibulocochlear nerve

8. Auditor tube

9. Tympanic cavity

10. Tympanic membrane

11. Cochlea

12. Eustachian tube

13. Malleus

14. Incus

15. Stapes

16. Auditory nerve

Interesting Fact

The inner ear contains the cochlea, a spiral-shaped organ filled with fluid and lined with sensory cells called hair cells. These cells convert mechanical vibrations into electrical signals, enabling us to perceive sound and detect head position and movement changes.

Ear

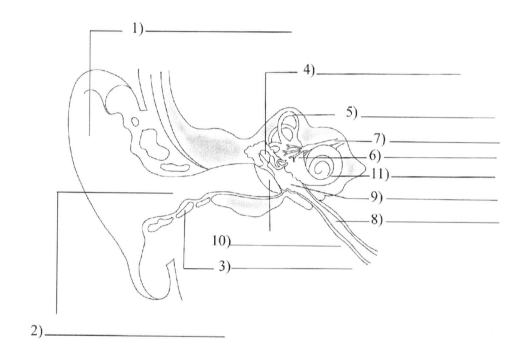

1) _____

4) _____

5) _____

7) _____

6) _____

11) _____

9) _____

8) _____

10) _____

3) _____

2) _____

The Nose

The nose is a vital organ of the respiratory system responsible for the sense of smell and the conditioning of inhaled air before it reaches the lungs. It consists of external and internal structures that play essential roles in olfaction, respiration, and facial aesthetics.

1. Root
2. Bridge
3. Dorsum nasi
4. Ala
5. Apex
6. Philtrum
7. Nasal bone
8. Septal cartilage
9. Major alar cartilage
10. Nasal vestibule
11. Nasal conchae
12. Olfactory epithelium

Interesting Fact

The nasal cavity is lined with specialized cells called olfactory receptors, which detect odors and transmit signals to the brain for interpretation. Humans can distinguish between thousands of different smells, showcasing the remarkable sensitivity and complexity of the olfactory system.

Nose

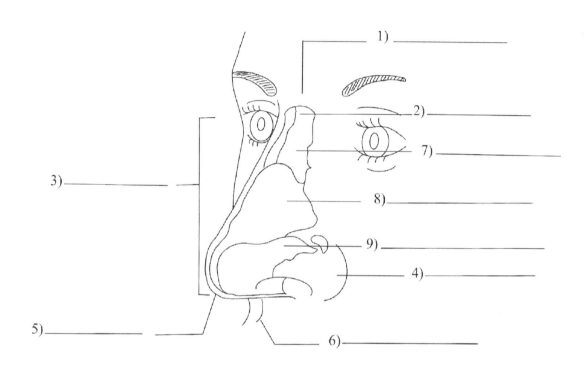

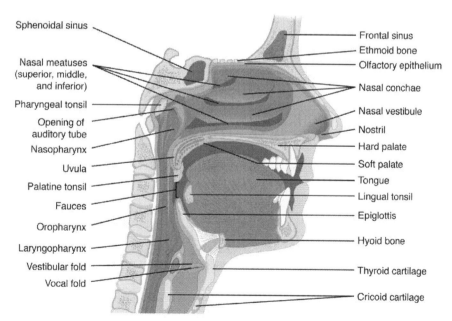

Conclusion

Congratulations on completing the Human Anatomy Coloring Book! By engaging in the coloring activities and studying the intricate structures of the human body, you have embarked on a journey of discovery and learning. Your journey through these pages, filled with intricate structures and fascinating details of the human body, marks a significant step in your quest for knowledge and understanding.

By engaging with the coloring activities and absorbing the information provided, you've enhanced your understanding of anatomy and cultivated valuable skills that will serve you well in your academic and professional endeavors.

We hope this coloring book has provided you with a fun and interactive way to explore the fascinating world of anatomy. Happy coloring, and may your curiosity lead you to insights beyond the pages of this book!

Made in the USA
Columbia, SC
14 November 2024